EUCLID'S FATHER BEARING HIS MOTHER ASHORE
ON HIS BACK

# EUCLID'S
# OUTLINE OF SEX

*A FREUDIAN STUDY*

BY

WILBUR D. BIRDWOOD

*Illustrations by Herb Roth*

NEW YORK
HENRY HOLT AND COMPANY

# PREFACE.

The present book has been written because it simply had to be written.

It is little short of amazing that earlier students of psychoanalysis should have so completely overlooked the extraordinary wealth of sex-material contained in the pages of Euclid, the Father of Geometry. It cannot be lack of interest, because nothing in life fails to interest the true Freudian. As the Roman poet has said, *Nihil humani a me alienum* — "There is nothing human that cannot be explained from the standpoint of the alienist." It cannot be lack of diligence, because it may be said of the true Freudian that he almost works overtime. It cannot be lack of information on the subject, because that has never been an insuperable obstacle.

Under the circumstances, I can only describe the Freudian neglect of Euclid as one of the most remarkable cases of inhibition on record.

And yet there is the fact. Euclid, to put it bluntly, reeks with sex. In no writer of ancient or modern times, with the possible exception of Legendre (*Elements of Geometry, Paris*, 1794), and Wentworth and Smith (*Plane Geometry, for the Use of High Schools*), does the theme of the Eternal Triangle run so persistently as in Euclid. Even the modern French drama succeeds in getting away from the triangle now and then; but Euclid, almost never.

Not that the Father of Geometry did not try. He took refuge in parallel lines, for example. But no sooner would he draw his parallel lines than Fate, or more properly, the Unconscious, compelled him to draw a third line cutting the

iii

parallels and so bring into existence triangles with their vertical angles equal. He resorted to quadrilaterals. And again the Unconscious would intervene and make him draw a diagonal, dividing the quadrilateral into two triangles. Whimsically he once referred to this sex diagonal as the bar sinister.

And so with circles, to which Euclid devoted himself in his later books. He was continually circumscribing the circle of life around the triangle of sex (See Figure 1), or inscribing the circle of life within the triangle of sex (See Figure 2). There was no conscious purpose in that. He would start off with his circle of life having its center at A (See Figure 3). He would move cheerfully along the radii of his circle to the circumference at B and C. And then, before he was aware of it, he had drawn a chord, BC, connecting the two radii and producing ABC — a triangle!

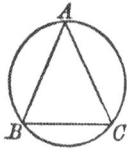 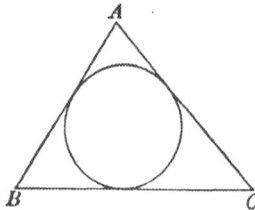 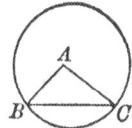

FIG. 1　　　　　FIG. 2　　　　　FIG. 3

I believe I have said enough to justify the present modest little volume, if modest is the right word. Profounder scholars and bolder thinkers are sure to follow where I have humbly shown the way. In due time, I have no doubt, the virtually inexhaustible sex-content of Euclid will have been made available for all students, with special emphasis on the Elementary and Kindergarten Grades.

W. P. B.

# CONTENTS

# ILLUSTRATIONS

## PART I

GRANDMOTHER

# CHAPTER I

## WHO WAS EUCLID?

The salient facts in the life of the great geometer, as recorded in the pre-Freudian text-books, are not hard to grasp.

Typical is the account of Euclid in the New International Encyclopedia (New York, 3rd edition, 24 volumes). There, in volume 8, page 153, under the title " Euclid," it is stated:

*Nothing is known of his life except that Proclus tells us that he lived in the time of Ptolemy I, who reigned* B.C. 306–283.

Quite as convincing is the sketch of Euclid's career in the Encyclopædia Britannica, 11th edition, volume 9, page 879. It is from the hand of John Sturgeon Mackay, M. A., LL.D., F. R. S. (Edin.), Chief Mathematical Master at Edinburgh Academy, 1873-1904; First President of the Edinburgh Mathematical Society; author of " Arithmetical Exercises " and " Elements of Euclid."

Professor Mackay tells the story of Euclid with all the art of a master of the miniature. He writes:

*We are ignorant not only of the dates of his birth and his death, but also of his parentage, his teachers, and the residence of his early years.*

Concerning Euclid's residence in his later years Professor Mackay intimates that we know nothing. The Britannica is an expensive publication, but even at two or three times the price, it could not have put the facts about Euclid's career more definitely.

## CHAPTER II
### WHO WAS EUCLID ? (*Continued*)

If the foregoing accounts of the life of Euclid should strike the reader as somewhat over-condensed, it is not the fault either of the New International Encyclopedia or the Encyclopædia Britannica. Given the methods of historical investigation prevalent at the time those two admirable works of reference were composed, the authors did the best they could.

No such limitation, however, rests upon the psychoanalytic student of the life of Euclid. All that is necessary is to apply to Euclid the method that has yielded such admirable results in Freudian studies of Leonardo da Vinci, of Goethe, of Mark Twain, of Margaret Fuller, and other notable figures. Immediately there emerges out of the cloud of suppressions and inhibitions which has gathered about the memory of Euclid an amiable and fascinating personality.

It is necessary only to fill in the sketch presented by the International and the Britannica, and the entire career of the great geometer rises vividly before us. Let us make the attempt.

# CHAPTER III

## HIS ORIGINS

We see Euclid born on the island of Kos in the early summer of the year 342 B.C. This fact makes it all the harder to understand why he should so frequently be confused with another Euclid, who was born in Sicily six hundred years earlier, and who attained fame as a wholesale cattle-dealer.

Euclid was born, then, on the island of Kos, and of a native mother, probably a member of the ruling family of the Delta Upsilons. Like most Greek gentlewomen of her time, she was brought up in complete ignorance of the arts of reading and writing, but trained in all the accomplishments of the complete housewife. She was an adept in the preparation of food, with the assistance of a professional female·cook named Myrtilla, who was probably of Thracian origin and had a special gift for fig pastes and stewed tripe in honey. Euclid's mother was also expert in the arts of spinning and weaving, under the supervision of another female slave named Sophronia, a native of Sparta. And naturally Euclid's mother was skillful in the care and feeding of children, under the guidance of her own old nurse Melanchthia.

In short, it can be said of Euclid's mother that she faithfully subscribed to the dictum laid down by Solon that woman's place is in the home.

The father of Euclid was a trader from Crete. On one of his voyages, presumably in the open winter of the year 344 B.C., he was shipwrecked on the coast of Kos, but succeeded in making his way to land carrying his aged mother on his shoulders.  This, of course, was the approved method of rescuing imperilled parents from fire, flood, and battle in old Greek times.

It may be asked whether it was the custom for traders in ancient Greece to take their old mothers with them on their voyages, especially as business men in those days frequently went in for piracy as a side line.  But as we shall see further on in this volume, the career of Euclid and the origin of the " Elements of Geometry " demand the existence of a paternal grandmother.  It follows, therefore, that Euclid senior did take his aged mother with him on his business trips, and that he did carry her to safety through the breakers on his shoulders.

## CHAPTER IV

### INFANCY AND CHILDHOOD

Euclid was the fifth of seven children, three boys and four girls, of whom the two youngest, both boys, did not survive infancy. The girls grew up and married well.

Euclid was fair-haired, big for his years, though with a slight tendency to stoop. He had a not unpleasant cast in his left eye, and up to the age of four was inclined to stammer. This last point will be found to be of vital significance when we come to consider the familiar Euclidian literary style as revealed in mannerisms like " Let angle *ABC* be equal to angle *DEF*," etc.

The boy's life was of more than normal happiness. It naturally would be. The study of Greek came easy to him. Latin, Modern History, Manual Training and, of course, Geometry, had not yet been invented.

Spelling and Composition were not difficult. The method of writing the Greek language was then in a state of transition. The conservatives clung to the fashion of writing from right to left. The liberal and radical classes went in for the newer method of writing from left to right. The boy Euclid wrote in either direction as the fancy seized him and it made no difference.

His schooling, as a whole, was spasmodic. Educational theory and methods were at that time in rapid flux. It was an age of intense experimentation. In the field of History, for example, there were several diverse methods of instruction. One method insisted on teaching dates without facts. Another method specialized in facts without dates.

7

A third method went pretty far in dispensing both with dates and facts and concentrated on Trends and Currents.

Euclid's mother, naturally ambitious for her son, sent the boy through all the available experimental schools, with the result that the boy attended eleven different educational establishments in less than a year and half. Is it any wonder that a boy so educated should have devoted his life to proving that a straight line is the shortest distance between two points?

# CHAPTER V

## DARK DAYS

When Euclid was six years old, his father perished in a raid upon the island of Kos by the Phi Beta Kappas, a pirate tribe inhabiting the adjoining mainland. His mother was carried off into captivity, but the lad and his grandmother were left behind as of doubtful commercial value.

Thus arose a Complex between the two which was immeasurably strengthened in the course of the next three years.

When the boy was nine years of age the old lady died, but not without leaving a profound impression on the future Proposition 16: "If one side of a triangle be produced, the exterior angle is greater than either of interior opposite angles." But this is to anticipate.

Concerning the attachment between the lad and his grandmother — altogether unnatural from the standpoint of psychoanalysis — the historian Archilongus has preserved the following legend:

To the end of his life — and Euclid lived to be seventy-six years, eight months and odd days, dying of sepsis following upon a badly ulcerated upper molar — the famous geometer devoutly observed the anniversary of his grandmother's death. On that day he would refuse to meet his students. He would bathe ceremonially, don a purple robe, and comb his beard with special care. He would partake of no food whatsoever, but, having sacrificed punctiliously to Hermes Mathematikos, he would give himself up to contemplation.

To his favorite pupil who questioned him on the subject,

Euclid explained that he devoted that day to evoking the memory of the aged woman who took him, a motherless child, to her own heart. She would go out every sundown into the olive groves to pick up kindling for a fire — they were very poor, Euclid told the disciple. Having given the boy his supper, she would croon him to sleep on her lap before the hearth.

# CHAPTER VI

## THE GRANDMOTHER COMPLEX

An exhibition of exaggerated sentiment, like that just described in the aged Euclid whenever he thought of his own orphaned childhood and the grandmother who nurtured him, might have been regarded by the ancient Greeks or the Chinese of today as piety. But Freud has taught us better. What the adult Euclid suffered from was unquestionably an aggravated case of the Grandmother Complex.

Here we find ourselves in need of further details about the life of Euclid. We experience no difficulty in deriving such additional data from the same sources that have served us so well in the preceding chapter.

Euclid first became actually aware of his grandmother when he was two years old. She had, of course, entered into her grandson's life before that. The old lady frequently differed with her daughter-in-law, Euclid's mother, on the subject of milk temperatures, pins, colic, talcum, relief for the gums, and other issues on which the old generation has always been in conflict with the middle generation. It resulted in the old lady's going to live with one of her married granddaughters, who as yet had no children of her own.

Nevertheless she continued to visit the Euclid household daily. When Euclid entered on his third year, a little brother was born, and the future mathematician was much more freely exposed to the unhampered influence of his grandmother. He came to look forward eagerly to her visits.

# CHAPTER VII

## WISH-FULFILMENTS

The reason was a simple one. Grandmother very seldom came to call upon the boy without a gift of some kind. Usually it was either a bagful of honey-cakes or a basket of dried sun-flower seeds, of both of which the little Euclid was inordinately fond. Sometimes the old lady would come in by the front door and sometimes she would use the back-door or the servants' entrance. The reader will do well to keep these two doors in mind.

Also this further fact: At times the little Euclid would immediately climb into his grandmother's lap and begin munching at the honey-cakes. At other times the two would go out strolling on the beach hand in hand. The old lady would point across the wine-colored waves to Crete where Euclid's father was born and had been, in his time, quite as tiny a tot as Euclid was now. The child listened and did not understand. But the honey-cakes and the sun-flower seeds were delicious.

Thus the Grandmother Complex was established in that young soul.

The Grandmother Complex

# CHAPTER VIII

## THE GRANDMOTHER TRIANGLE

We are now in a position to grasp the real meaning of what is probably the best known of Euclid's later contributions to the sum of human knowledge.

*If two triangles have two sides of the one equal to two sides of the other, each to each, and have also the angles contained by those sides equal to one another, they also have their bases or third sides equal; and the two triangles are equal; and their other angles are equal; each to each, namely, those to which the equal sides are opposite.*

Euclid's demonstration is a model of condensed, if somewhat dictatorial, literary expression. He says, virtually —

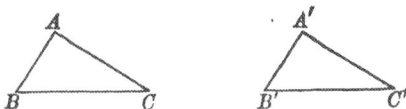

FIG. 4

In the above triangles let the line $AB$ be equal to the line $A'B'$, and the line $AC$ to the line $A'C'$, and the angle $BAC$ to the angle $B'A'C$; then will the line $BC$ be equal to line $B'C'$, and the two triangles will be equal in every respect.

For, superimpose the second triangle on the first. Then will the line $A'B'$ coincide with $AB$ and point $B'$ will fall on point $B$. But since the angle $B'A'C'$ is equal to the angle $BAC$, the line $A'C'$ will take the direction of the line $AC$, and point $C'$ will fall on point $C$.

Now, if point $B'$ coincides with point $B$, and point $C'$ with point $C$, the line $B'C'$ will coincide with the line $BC$, and the two triangles are equal in every respect. Which was to be demonstrated; popularly referred to as $Q.E.D.$

# CHAPTER IX

## DOUBTS

This is the way the pre-Freudian text-books in Geometry used to present Euclid. Superficially the demonstration was adequate. Actually it was devoid of meaning. All sorts of doubts were bound to arise.

For example: if the triangles $ABC$ and $A'B'C'$ are indeed equal in every respect, why bother with two triangles? Life is short enough as it is. This doubt is one that has perplexed generations of high-school students. They have wondered why for all practical purposes the world could not get along with $ABC$ only. They had their moments, indeed, when they felt that they could very well get along without any triangles at all; especially at the beginning of the baseball season.

All through Euclid we find him at great pains to demonstrate, by roundabout ways, truths that were either obvious, or unimportant, or both. Euclid, for example, lays great stress upon the fact that in any triangle any one side is shorter than the sum of the other two sides. Yet this truth has always been self-evident to a small boy with a bigger boy after him; to a dog with a fire-cracker attached to his tail; or to a commuter bound for the 7:34.

It is like trying to prove, by mathematics, that you can get a drink in Havana.

# CHAPTER X.

## THE REAL ˌSOLUTION

Such doubts are inevitable if we persist in reading Euclid in the old manner. But how if we bring psychoanalysis to bear on the problems?

The triangles with which Euclid was so greatly concerned in his later life were not plane triangles but sex triangles.

In the theorem discussed in the preceding chapter, the triangle $ABC$ really represents the infant Euclid's exaggerated emotional reaction to his grandmother. The triangle $A'B'C'$ is the later sex-expression of that Grandmother Complex.

In the infant triangle $ABC$, point $A$ would be the child Euclid. He catches sight of his Grandmother coming in with the honey-cakes at the front-door, $B$, or with the sunflower seeds at the side-entrance, $C$. Then the line $BC$ would be the locus or track of the child's appetite. Instead of the line $BC$ we might quite as well say Honey-cake-Sunflower line.

What follows is simple. In the adult sex-triangle $A'B'C'$, our Euclid sets out from the same point, $A'$, himself. He goes on thinking along the line $A'B'$ until the ancient inhibition brings him to a stop at B', the honey-cakes. Or if he starts out in another direction, the permanent angle given to his infant soul by his grandmother, impels him along the line $A'C'$ till the same inhibition brings him to a stop at point $C'$, the sunflower seeds.

Thus the line, $B'C'$, representing the sex life of a mature scientist, is predetermined along the old Honey-cake-Sunflower line, $BC$.

Euclid, of course, thought he was inventing Geometry. Actually he was rehearsing a vivid anxiety-dream and wish-fulfillment of his childhood. Or if we insist on calling it Geometry, the very least we can do is to speak of it, not as Plane Geometry, but as Complex Geometry.

This Complex runs all through Euclid; so much we can say by way of anticipation. Wherever he demonstrates that $ABCDXWJZ$ is equal in every respect to $A'B'C'D'W'X'J'Z'$, we are only in the presence of a phenomenon technically described, for obvious reasons, as the Przemysl Complex.

Whenever Euclid asserts that two triangles are equal each to each, or two quadrilaterals are equal each to each, or two circles are equal in all respects, it is only Euclid's way of saying that it is sex of one and half a dozen of the other.

# CHAPTER XI.

## PROBLEMS INVOLVING THE GRANDMOTHER COMPLEX

*Parallel lines are always equidistant from each other.*

A —————————————— B
C —————————————— D

In the above parallels, *AB* represents Euclid, and *CD* represents his Grandmother, walking hand in hand.

*Parallel lines cut by a third transverse line produce internal and external angles.*

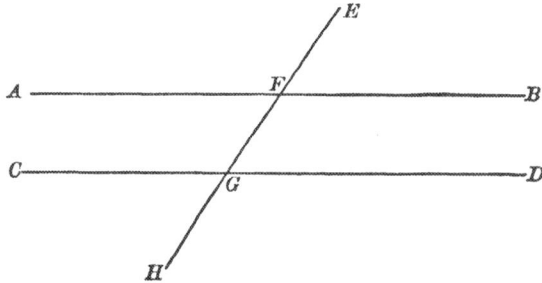

Fig. 5

In the above diagram *AFE* is an external or extraverted angle. It represents Euclid watching his Grandmother approach with honey-cakes. The angle *AFG* is an internal or introverted angle. It represents Euclid anticipating the feel of the honey-cakes inside of him.

18

*Every point in the circumference of a circle is equidistant from the centre.*

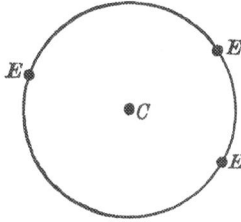

FIG. 6

In the above diagram the center $C$ is Grandmother on the rare occasions when she forgot to bring the honey-cakes. The circumference $EEE$ is Euclid giving Grandmother a thorough examination from every point of view and hoping against hope.

Note: The circumference of a circle is equal to 3 1/7 times the diameter; at least, in the very best circles.

*A straight line is the shortest distance between two points.*

$A$ ————————————— $B$

In the above diagram $A$ is Euclid and $B$ is Grandmother, and the line $AB$ represents Euclid making a bee line for Grandmother, this being the shortest distance between two points.

*Things equal to the same thing are equal to each other.*

Example: The Eighteenth Amendment can be explained only in terms of sex, and Michelangelo is explainable only in terms of sex. Therefore, the Eighteenth Amendment and Michelangelo are identical.

# PART II

OEDIPUS

# CHAPTER XII

## WHO WAS OEDIPUS?

Advanced students of psychoanalysis have, of course, recognized for a good many pages back the true nature of the Grandmother Complex. They have understood that the Grandmother Complex is only another form of the Oedipus Complex; only much more so.

For the benefit of beginners in psychoanalysis it may be permitted to explain just what is this Oedipus Complex, in a few brief but vivid words.

The Oedipus Complex is the neural mix-up which accounts for the familiar fact that children are more strongly attracted to the parent of the opposite sex. The son is fonder of his mother; the daughter is more strongly drawn to her father.

But to say fonder is to put it mildly. At bottom every Freudian knows that every normal son would like to see his father dead and every normal daughter would like to see her mother dead. This pleasant truth has been established by the special researches of Freud's famous collaborator, Friedrich Jung. Hence the expression, the Jung Generation.

The name for this charming little Complex comes, of course, from Oedipus. He is known in Greek legend as the hero of three exceptionally difficult feats: he slew his own father; he married his own mother (both in ignorance, to be sure); and he answered the riddle of the Sphinx, which no one before him had been able to solve.

The story of Oedipus was utilized by the Greek dramatists for a number of tragedies. The Greeks called them tragedies because the Greeks lived in the days before Freud and were the slaves of certain established prejudices concerning the proper relations between parents and children. Today we would call these plays Social Comedies.

# CHAPTER XIII

## THE OEDIPUS IN ALL OF US

The basic truth of the Oedipus Complex is confirmed by everyday experience. Approximately one-half of my readers, for example, must at some time have been little boys. Such readers will have not the least difficulty in recalling their experiences with measles or scarlet fever. Upon such occasions, when they were particularly feverish or otherwise ill-at-ease, they almost invariably called for Mother. It is also a well-authenticated fact that wounded soldiers in their delirium most frequently call for their mothers.

On the other hand, those of my readers who were once little girls, will recall that on occasions during the summer vacation when they were confronted by a strange dog, or an infuriated hen, or a cow whose intentions were doubtful, they almost invariably called for Daddy.

The author may be pardoned for citing an instance of the Oedipus Complex from his own experience. He will never forget how during his bitter struggles with the first book of Euclid, it was almost invariably his father who insisted that the boy finish his prescribed lessons before he went to bed, whereas his mother would argue that the boy had had a hard day, and it was more important for him to be in bed by nine o'clock, not forgetting to brush his teeth before retiring.

There were other times when the boy would turn to his father for assistance in arriving at the truth that from any point outside of a line only one perpendicular can be drawn to that line. His father would begin buoyantly enough.

He would study the text for a minute or two. Then his face would grow stern and he would remind the boy roughly that it was dishonest for a school boy to solicit help in preparing his home-work.

His mother, on the contrary, would either help him out with the perpendicular or else say that it was time to go to bed, not omitting the tooth-brush. As a result, all through high-school the author recalls that his feelings for his mother were much more tender than for his father.

To be sure, it did not go so far as the boy's wishing that his father were dead. But he did have dreams in which his father used to ride an old-fashioned bicycle on a tight-rope, condemned to solve innumerable problems about a perpendicular from a point in the handle-bar to said tight-rope.

## CHAPTER XIV

### THE ANTI-OEDIPUS DIE HARDS

The Oedipus Complex has not been without its critics. They argue that frequently a little boy when pursued by a mad hen, will call for Daddy instead of Mother. They cite the case of little girls who, in dangerous proximity to a strange cow, will call for Mamma. But such examples must be dismissed as irrelevant.

Neither is there much force in the contention that sons are more attached to their mothers because of a protective male instinct, and daughters are more attached to their fathers because of a feminine mother-instinct. Everybody has been saying this for thousands of years and it therefore cannot be true.

The only real difficulty we encounter is in the case of young children, who are asked by visitors whom they like better, Mother or Daddy, and who reply, " I like Mamma and Daddy best." The explanation of this will most likely be found in a super-Oedipus Complex.

# CHAPTER XV

## THE RIDDLE OF THE UNIVERSE

One apparent omission we do encounter in the handling of the Oedipus Complex by the psychoanalysts. They never touch upon the story of Oedipus and the riddle of the Sphinx.

It will be recalled that the Sphinx asked Oedipus: What animal walks on four feet in the morning, on two feet at noon, and on three feet at night. Oedipus replied that the animal is Man; he crawls on all fours in infancy, walks erect in manhood, and leans upon a staff in old age.

Up to recently this answer may have been deemed sound enough; but now we see it to be false. For if Freudianism has taught us anything, it has taught us that man is an animal who never walks erect, but is always crawling around in the mud.

As a matter of fact, psychoanalysis does concern itself with the riddle of the Sphinx without mentioning it by name; and with a certain difference.

In the old Greek legend, Oedipus discovered the answer to the riddle.

In our own psychoanalytic science Oedipus is himself the answer.

What is life? Oedipus Complex. What is thought? Oedipus Complex. What is health, disease, love, hate, cruelty, martyrdom? Oedipus. What is art, literature, war, the pulpit, the professor's chair, the bacteriologist's microscope? Oedipus. Why does the writer write, the engineer build, the mathematician draw figures, the poet sing, the subway guard howl, the conservative conserve, the radical protest, the middle-of-the-road man stay in the middle? Oedipus.

Implanted in man is this passion for finding a single answer to the riddle of everything. At different times the answer has been providence, fate, chance, fire, water, air, earth, the atom, the electron, the amoeba. Often it has been ouija, the violet ray, internal bathing, the Kneipp cure, mineral oil, bran, Fletcherism, the bicycle, the automobile, the Soviet, the simple life, the daily dozen, the solar plexus. Always it has been a single magic word to satisfy all of the Sphinx's questionings.

Today Freud has given us the real answer, the all-sufficient answer. What is anything, and why, and how? Ans. Oedipus.

# PART III

BALAAM

# CHAPTER XVI

## THE ASS AT THE BRIDGE

I believe I have by this time convinced the open-minded reader that there is hardly a page in the " Elements of Geometry " that will not yield rich treasures of sex meaning to the searcher equipped with the proper psychoanalytical tools. But I have reserved, until this moment, consideration of one particular Proposition in the " Elements " which seems to me to sum up the entire Freudian gospel as it was anticipated and exemplified by Euclid.

This Proposition occurs very early in Book I, and is usually stated as follows:

*If two sides of a triangle are equal, then the angles opposite these sides are equal.*

In the history of mathematics this celebrated proposition has come to be known as the Pons Asinorum, or the Bridge of Asses.

The phrase is not altogether a happy one. The reader at first sight would assume that a Bridge of Asses is a bridge peculiarly adapted for crossing by an Ass. Whereas Euclid's intended meaning is just the opposite. It is a bridge that an Ass cannot cross, try how he may.

The common explanation of the Pons Asinorum is that this is the Proposition where the slow-witted student of Euclid is brought to a standstill. Hitherto he has managed to stumble along with more or less difficulty. But when confronted with a triangle with two equal sides the dunce shies, balks, and refuses to cross. Either he transfers from Geometry to Comparative Literature, or else he changes

31

to a new school where emphasis is laid on Character Development.

With all due modesty, however, I am compelled to state that this interpretation of the Bridge of Asses, so long endorsed by greater scholars than myself, is nevertheless to be rejected out of hand. To accept the common explanation would be to take the very heart of sex out of Euclid, as I shall now proceed to show.

Scholars and mathematicians would never have been content to regard the Bridge of Asses as a mere humorous epithet for the stupid school boy, if they had only stopped to consider one simple fact: In all literature and in all common speech both Ass and Bridge are words steeped in sex significance. They are erotic symbols *par excellence,* or as we should put it in English, pure and simple.

## CHAPTER XVII

### BRIDGES

The Bridge as a symbol of eroticism is so obvious and so familiar as to call only for the briefest consideration.

The Bridge is plainly the transition period from childhood to adolescence. That period corresponds with the eighth grade in the elementary schools and the first year of high school. In the study of mathematics it is the period when the student passes from arithmetical problems in carpeting floors and papering rooms at so much a square yard (excluding the windows), to the first principles of Euclid.

The Pons Asinorum would thus come very near the period when early adolescence is filled with the vague yearnings and perplexities to which psychoanalysis has given us the key.

This controlling sex connotation I discern without difficulty in "The Bridge of Sighs," and "I Stood on the Bridge at Midnight," which mistaken parents are accustomed to regard as an affliction when recited by their offspring, instead of recognizing them and welcoming them as the expression of a vital urge. But I am not quite prepared to go with some eminent members of the Left Wing of Psychoanalysis who discern the same sex-significance in "London Bridge is Falling Down."

There can be no doubt, however, about the real meaning of other famous Bridges.

Horatius at the Bridge, defending the young Roman civilization against the much older civilization of the Etruscans, is really a sex inhibition or a sex transition, as we may prefer to regard it.

The great Natural Bridge in Rockbridge County, Virginia, is just what its name implies, natural; that is to say, sexual.

The shot fired by the embattled American farmers at the Bridge at Concord, Mass., in 1775, is the exact parallel to Horatius. Emerson knew what he was talking about when he described it as the shot heard round the world: it was sex explosion.

The four bridges which span the East River, namely, Brooklyn Bridge, Manhattan Bridge, Williamsburg Bridge and Queensborough Bridge, connect the two most important sections, that is to say, sexions, of Greater New York.

A separate chapter might be written about the Bridge in Dentistry, both in itself and in relation to the Crown and the Root. But time presses.

## CHAPTER XVIII

### EQUUS ASINUS CYPRIENSIS

Concerning the Ass in general or as the philosophers might say, the *Ass sub specie aeternitatis*, it is necessary to repeat only what I said a little while ago.

This familiar animal has in all ages and all climes been a symbol of eroticism, together with the Bird, the Cat, the Donkey, the Eagle, the Fur-bearing Seal, the Giraffe, the Hyena, the Irrawady Woodpecker, the Jaguar, the Kangaroo, the Llama, the Mesopotamian Fish-Hawk, the Narghili, the Ox, the Penguin, the Quadriga, the Rhinoceros, the Swan, the Tourniquet, the Uganda, the Vituperative Buzzard, the Weasel, the Xingu, the Yuban, and the Zebra.

It remained for the Romans to coin a phrase about the Ass, descriptive of the fact that by means of erotic symbols we can attain the truth that lies beyond the stars: *Per Asspera ad Asstra.*

## CHAPTER XIX

### BALAAM

We may now pass on to an examination of a few of the most celebrated erotic Asses in history.

The field is both extensive and crowded, and anything like an exhaustive study is, of course, out of the question. Our investigation may well begin with the earliest specimen on record which fortunately happens to be also the most typical and famous example of the species.

This is, of course, Balaam's Ass, as we first encounter him in Numbers, XXII, 22–33, and subsequently in all world literature.

The story of Balaam and his Ass may be briefly rehearsed for the benefit of those readers whose restricted leisure has not brought them into contact with the ancient literature of the Hebrew people.

When the Israelites, on their way from Egypt towards the promised land, reached the frontiers of Balak, King of Moab, that ruler was naturally alarmed by the overwhelming strength of the invaders. He therefore sent messengers in great haste to the celebrated wizard Balaam, with the urgent request that Balaam hasten to the court of Balak and curse the Israelites for him.

Balaam, after some delay marked by a number of vivid dreams, inhibitions, and frustrated wish-fulfilments, consented to undertake the journey to Balak's court, and mounted his Ass for that purpose.

What follows is pure Freudianism. The Ass immediately became aware of the presence of an angel blocking the road.

36

Balaam and the Ass

Balaam, on the other hand, remained in dangerous ignorance of the obstacle.

Difficulties ensued. First the Ass turned aside into a field. Then he crushed Balaam's foot against the wall of the abutting vineyard. Then he fell prostrate in the road. Meanwhile Balaam was cruelly engaged in repressing the Ass's desires with his heavy staff until the inevitable neurotic discharge occurred; the mouth of the Ass was opened and he addressed his master in a few well-chosen words with the precise tenor of which we are not particularly concerned.

The significant fact is that the Ass did break into speech.

## CHAPTER XX

### THE ASS AND THE UNCONSCIOUS

" The significant fact is that the Ass did break into speech."
For stressing this point the author of the present volume
has at least two compelling reasons. One may be called
personal. His own emotional reactions to the vocal Ass in
the Balaam epic are so strong that to abstain from utterance
on the subject would be to practice inhibition of a most dangerous kind. Feeling as I do about Balaam, it must be
either the fullest and frankest kind of speech or else an
inevitable nervous breakdown.

The second reason has to do with the public and with its
right to the Truth. It is my sincere conviction that I have
discovered the Truth about Balaam's Ass, and to withhold
that Truth from the world at large would be a social crime.
It is every man's inescapable duty to give the public what
it wants, even if it wants the Truth, as it occasionally does.

38

# CHAPTER XXI

## SUBLIMINAL

Consider, then, the two basic elements in this Balaam-Ass epos: the fact that it involves an Ass and the fact that the Ass is represented as speaking for publication.

On the first point I have possibly said enough when I have shown that the Ass is in himself a pure erotic symbol. Whenever in history or legend or poetic fantasy we come across the Ass as a central figure, we may feel confident that it is a case of *Cherchez la femme*.

But why is the Ass, aside from his innate eroticism, so splendid an embodiment of the Freudian idea? The answer is simple. The Ass is also the perfect type of the Unconscious.

The Ass is the traditionally patient beast of burden. He carries his rider on his back precisely as the Unconscious carries the real burden of our conscious mental life. The Ass submits to the rein, the whip, the staff, the spur, as though he were dependent for his very existence on the good will of his master. So the Unconscious submits to checks and controls as though it existed by sufferance of the Intelligence, instead of actually carrying the Intelligence on its back, so to speak.

But, when the need arises, the Ass, like the Unconscious, will balk, rear, kick out, and in the extreme, send his rider head foremost to the ground. In the final test the Ass and the Unconscious have little difficulty in telling the rider, whether it be Man or Intelligence, just where he gets off.

And now consider Balaam and the contrast. The Ass is

39

supposedly a stupid, reticent, patient beast of burden, bound down to humdrum duties. Balaam, on the other hand, is a wizard, a seer, a something of a prophet. He has the gift of eloquence. He has the power to bless and curse. The legend in Numbers XXII represents him even as in communication with God. . . .

And yet when it comes to the ultimate test between primitive instinct and the Unconscious, as typified in the Ass, and conscious pride of Intellect, as typified in Balaam, it is the Ass who wins out. It is Balaam who gets his foot crushed against the wall, and is dragged through the brambles, and ends by biting the dust. And the lesson is reinforced by a few illuminating remarks from the hitherto inarticulate Ass.

And so it always has been in the last resort. Whenever the tyranny of the Intelligence becomes unendurable, the Ass grows vocal.

# CHAPTER XXII

## SPEECH

What vernacular did the Ass employ in addressing the prophet Balaam? Presumably the Hebrayic.

The bray of the Ass has a poignant symbolism that far outstrips in significance its physical carrying power. It is penetrating in something more than the ordinary sense. It penetrates into the deeper meaning of life. The Ass is vocal but not local. He is universal.

The Ass's bray is universal precisely for the reason that it is monotonous. Like the Unconscious, the Ass expresses himself in only one key, and for the very simple reason that there is only one key to the mystery of life — Unconscious Sex.

Compare the splendid universal monotone of the Unconscious Ass, with the confused diversity of the human voice. It is the difference between a real world-language and Babel.

The human voice has developed an extraordinary variety in its attempt to express the supposed variety of the Scheme of Things. It has developed rhythm, tone, pitch, timbre, modulation, inflection. It has developed scales and gamuts. It makes use of sharps, flats and naturals. It has sought refinement of expression through breves, semibreves, crotchets, quavers and demisemiquavers. It has even invented that mysterious thing called " temperament " which is supposed to add a new meaning to the normal effect of the physical sound vibrations.

And it is all vanity.

The human voice believes that it has fitted the **proper**

intonation to the proper sentiment. It has one tone for love, another for pity, another for terror, another for mercy, another for affection, another for self-sacrifice, another for cruelty, another for courage, another for cowardice. And it proceeds on the pathetic assumption that when it utters these diversified sounds it is mirroring a corresponding reality.

The human voice actually believes that there are different things called love, pity, terror, mercy, affection, self-sacrifice, cruelty, courage, cowardice.

The Ass knows better. He has only one tone because he knows that there is only one Reality, and its name is Venus Cytherea.

The human voice sings, shouts, weeps, lilts, groans, ejaculates, cheers, ululates, lullabies, chants, curses and exhorts, under the impression that it is responding to different stimuli.

The Ass knows there is only one stimulus, and it is the Cyprian Aphrodite, or to give her the older name, Astarte, that is to say, Asstarte.

So he brays.

The human voice, in its quest for truth, stutters, whispers to itself, encourages itself, scolds itself, gasps with awe, breaks into babble with the joy of understanding, says " I have it! " says " Goldarn it! " spends itself, in other words, upon shadows and appearances.

The Ass knows better. He does not whistle when things are going fine under the lens of his microscope. He does not ejaculate when something new swims within the field of his telescope. He does not groan when things go wrong on his electromagnetic weighing-scales. He does not shout when things begin to precipitate in his bacteriological test tube. He does not swear when the two halves of his new iron bridge run out of true by a quarter of an inch. He does not halloo when the two halves of his Vehicular Tunnel meet within half an inch. The Ass knows that there is only one answer. It is She of the groves and the High Places — with complications. It is Cybele, Oedipus, Aphrodite, Narcissus, and the Maenads. It is all in the family.

So he brays.

The human voice tries to utter its sense of the manifoldness of life, and it only succeeds in developing the complicated symptoms described in the programme notes of the Philharmonic Symphony Orchestra.

The Ass asserts the essential sex-unity of the universe. He brays.

# CHAPTER XXIII

## SUMMARY

Before we say farewell to Balaam it will be of service to restate the significance of this Freudian epos. The story of Balaam is the story of sex. The evidence may be marshalled in a few brief statements.

The King of Moab is terrorized by an invading host of Israelites coming from the west. But we know that the West has always been a symbol of sex.

The King of Moab summons to his aid a wizard from the east. But the East has always carried a peculiar sex-significance.

· The King of Moab in his letter to Balaam doubtless referred to the invaders as like unto a thundercloud from the north. The North has always had an erotic significance.

The King of Moab, in this same letter, probably described the invaders as like unto a host of grasshoppers from the south. But the South can be said to be almost synonymous with sex.

The Israelites, fleeing before their Egyptian foes, crossed a desert. All Freudians know what a desert means.

The Israelites also encamped at the foot of a sacred mountain. All Freudians know what a mountain means, especially a sacred one.

The Israelites at various times came into contact with cloven rocks, burning bushes, golden calves, arks, tablets, inspired men, inspired women, traitors, rebels, sinners, saints, manna from the heavens, grapes from Eshcol, rivers, oases, tents, droves, Amalekites, Peruzzites, Jebusites, etc. But

44

to any one who knows what's what in Freud, it is hardly
necessary to point out just what is meant by cloven rocks,
burning bushes, golden calves, arks, tablets, saints, sinners,
manna, grapes, rivers, droves, Jebusites and Peruzzites.

When Balak, King of Moab, felt impelled to curse the
Israelites, it is clear that he suffered from an anti-Israelite
Complex.

When Balak was afraid to tackle the job himself but sent
for Balaam to do the cursing, he was plainly suffering from
an Inferiority Complex.

When Balak insisted on Balaam's coming, in spite of the
latter's refusal, he was plainly the victim of a Wish-fulfil-
ment.

When Balaam refused to come the first time, and hesi-
tated the second time, he was plainly suffering from a bad
case of Inhibition.

When Balaam sought his heart for counsel, he was obvi-
ously the victim of a bad case of Introversion.

When Balaam belabored the Ass with his staff and found
himself in the ditch as a result, he was just as plainly a
victim of Extraversion.

When Balaam lifted up his voice and said, " How beautiful
are thy tents, O Israel," he was giving utterance to an Incest
Complex.

The Ass was the only actor in the whole affair who was
immune from inhibition, repression or complex. He spoke
right out in meeting.

# CHAPTER XXIV

## PARENTHESIS

In passing, I cannot refrain from voicing my astonishment at the fact that students of Freud have hitherto given so little attention to the Bible.

Prejudice must not blind us to the fact that the book is simply crammed with the raw material of psychoanalysis. The few humble results I have derived from the story of Balaam are an index of what might be done by others with more time on their hands. I have but scratched the surface. The Song of Songs alone ought to be worth a million dollars to any trained Freudian scholar and practitioner.

No other book can touch the Bible for its wealth of material on sleep, visions, and dreams. Though a chronic sufferer from insomnia, I can testify to having repeatedly put myself to sleep by repeating all the generations from Adam to Zerubbabel.

Sleep, dreams, visions, hysterias, ecstasies, inhibitions, commandments, complexes, fulfilments, frustrations — the list is virtually inexhaustible. I will content myself with only one point which demonstrates the sympathetic attitude of the Book as a whole to the psychoanalytic world-outlook.

This is the emphatic Biblical assertion, that the ideal span of man's life is three score years and ten.

It is only another way of saying when a man has finished being a sexagenarian, life has very little further meaning.

PART IV

BURIDAN

# THE FORDNEY COMPLEX

**Par. 25.** Coal-tar products: Acetanilide not suitable for medicinal use, alpha-naphthol, aminobenzoic acid, aminonaphthol, aminophenetole, aminophenol, aminosalicylic acid, aminoanthraquinone, aniline oil, aniline salt, anthraquinone, arsanilic acid, benzaldehyde not suitable for medicinal use, benzal chloride, benzanthrone, benzidine, benzidine sulfate, benzoic acid not suitable for medicinal use, benzoquinone, benzoyl chloride, benzyl chloride, benzylethylaniline, beta-naphthol not suitable for medicinal use, bromobenzene, chlorobenzene, chlorophthalic acid, cinnamic acid, cumidine, dehydrothiotoluidine, diaminostilbene, dianisidine, dichlorophthalic acid, dimethylaniline, dimethylaminophenol, dimethylphenylbenzylammonium hydroxide, dimethylphenylenediamine dinitrobenzene, dinitrochlorobenzene, dinitronaphthalene, dinitrophenol, dinitrotoluene, dihydroxynaphthalene, diphenylamine, hydroxyphenylardinitrotoluene, dihydroxynaphthalene, diphenylamine, hydroxyphenylarsinic acid, metanilic acid, methylathraquinone, naphthylamine, naphthylenediamine, nitroaniline, nitroanthraquinone, nitrobenzaldehyde, nitrobenzene, nitronaphthalene, nitrophenol, nitrophenylenediamine, nitrosodimethylaniline, nitrotoluene, nitrotoluylenediamine, phenol, phenylenediamine, phenylhydrazine, phenylnaphthylamine, phenylglycine, phthalimide, quinaldine, quinoline, resorcinol not suitable for medicinal use, salicylic acid and its salts not suitable for medicinal use, sulfanilic acid, thiocarbanilide, thiosalicylic acid, tetrachlorophthalic acid, tetramethyldiaminobenzophenone, tetramethyldiaminodiphenylmethane.

# CHAPTER XXV

## OLD FRIENDS

Just about six hundred years ago there lived in France a philosopher and logician by the name of Jean Buridan. His claim to a place in history and to a portion of the strictly limited space at my disposal may be said to rest upon the back of no less than two Asses. Buridan must thus be regarded as the outstanding eroticist of the Middle Ages.

Of the two Buridan examples of the species *Asinus*, one animal is by an extraordinary coincidence — but I personally think it more than a coincidence — our old Euclidean friend the Ass at the Bridge.

Buridan, too, had his Pons Asinorum, but he made use of him in the field of logic and not of mathematics.

My readers are no doubt aware that in Logic they have a form of reasoning, or, if you wish to call it so, a form of entertainment, known as the Syllogism. The Syllogism is composed of a major term, a middle term, and a minor term, and the idea is to cross from the major term to the minor term by means of the middle term.

Now, there are syllogisms which are so simple that frequently there is no need to state the middle term; the ordinary logician can leap straight across from major to minor. Buridan, however, always insisted upon inserting the middle term, and it is supposed that he did it for the convenience of the Grade Ass student in Logic.

There is thus a notable difference between Euclid's Pons Asinorum and Buridan's Pons Asinorum. Euclid's Bridge of Asses was one which an Ass could not cross. (See page 31.)    Buridan's Bridge was built especially for the facilitation of Ass traffic.   Euclid, in other words, believed in the inhibition of Asses, while Buridan believed in the liberation of Asses.   But the important thing is that they both recognized the significance of the animal.

# CHAPTER XXVI

## SELF-DETERMINATION

However, the Buridan Ass who crossed the bridge is by far the less celebrated of his two specimens. The Buridan animal that enjoys universal fame is the Ass of the Haystacks.

Jean Buridan in philosophy was what is called a Determinist. That is to say, he believed that man's conduct is entirely guided and determined by external forces. To illustrate this point Buridan imagined an Ass placed at a precisely equal distance between two precisely equally attractive haystacks. It was Buridan's contention that an Ass, so conceived and so dedicated, would find it impossible to make up his mind between the two piles of hay and would slowly but surely perish of starvation.

The true meaning of Buridan's Ass is, of course, just the reverse of this commonly accepted version. Buridan did not really mean that the Ass, or Man, was bound to be permanently inhibited. He was merely anxious to point out the dreadful results of inhibition if carried too far. By the Ass starving to death he did not mean physical death, but psychic breakdown, neurosis.

Buridan, if he were living today, would have been all in favor of psychoanalyzing the Ass. He would discover the particular Complex developed by the Ass in early foalhood, bring that Complex to the surface of consciousness, and thus cast it out.

Thereupon the Ass would head without loss of time for one of the two haystacks. Having consumed it, he would turn his attention to the other haystack.

THE BROADWAY COMPLEX
The Messrs. Lee and J. J. Shubert Present

## CECIL LEAN AND CLEO MAYFIELD

IN

## "THE BLUSHING BRIDE"

Books and Lyrics by Cyrus Wood
Based on a Play by Edward Clark and Mark Swan

Music by Sigmund Romberg
Staged by Frank Smithers
Musical numbers arranged by Jack Mason
Orchestra Under the Direction of Al Goodman
Scenes by Watson Barratt

The Entire Production Under the Personal Direction of Mr. J. J. Shubert
Miss Lorraine's Gowns by Madame Frances

The sable and ermine wrap worn by Miss Lorraine in the Third Act
executed by C. C. Shayne and Co.

All gowns and costumes by Anna Spencer, Inc., Designs of Shirkey
Braker, Personal Supervision of Miss Anna Spencer

Boy clothes by Knox                     Uniforms from Brooks Co.
Shoes from I. Miller

Scenery designed and executed by Clifford Pember. Painted by Berg-
man. Built at Selwyn Studio.

Motion pictures by Shackleford.

It is asserted by the pre-Freudian text-books that there is really no mention in Buridan's extant works of either the Ass at the Bridge or the Ass of the Haystacks, and it is suggested that both Asses were, so to speak, saddled upon him by his enemies. But from the standpoint of the Freudian analysis the mere fact that a thing is not so does not prove that it isn't so. We shall be safe in continuing to think of Buridan as being himself the mediaeval bridge between the ancient eroticism of Euclid and the eroticism of today.

We may call Buridan, in fact, the Pons Asinorum between the classic and the modern world.

# PART V

THE NEW ENGLAND ELEMENT IN EUCLID

## CHAPTER XXVII

### THE EUCLIDEAN STYLE

All of us are acquainted with the phenomenon of what the older psychology used to call diffidence or shyness. (See W. S. Gilbert, *The Mikado*, Act II.) We know today that shyness in adults is the result of inhibition or repression in childhood.

The stigmata of the diffident man are familiar. In company he either stutters or yells. Either it is impossible to get him to talk or else it is impossible to choke him off. He is alternately panic-stricken or terrifyingly Napoleonic. When he is silent it is the Inferiority Complex. When he refuses to be flagged it is the Defense Complex. The louder he talks the more he is on the defensive.

Thus it was with Euclid. Remember his childhood of suppressions. Remember his enslavement to the Honeycake Complex and the Sunflower Complex. You will then understand the peculiar literary style of the Euclid we meet in the " Elements."

It is a hard-bitten, dogmatic, finger-pointing, bumptious style, with its " Let $A$ and $C$ be the two points," its abrupt " Draw a line from $F$ to $G$," its " Now's " and " Then's " and " Therefores," its self-satisfied, smug, almost smirking " *Q.E.D.'s.*"

We find Euclid at his characteristic best, or worst, in the following:

THEOREM

*A straight line perpendicular to one of two parallel planes is perpendicular to the other also.*

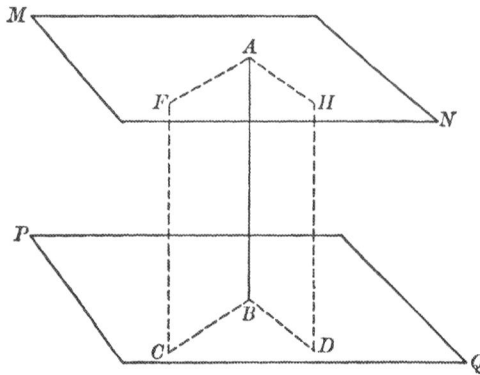

FIG. 7

Given the plane $MN \parallel$ plane $PQ$, and $AB \perp PQ$.

To prove      $AB \perp MN$.

Proof. Through $AB$ pass a plane intersecting $PQ$ and $MN$ in the lines $BC$ and $AF$ respectively; also through $AB$ pass another plane intersecting $PQ$ and $MN$ in $BD$ and $AH$, respectively.

| | | |
|---|---|---|
| Then | $BC \parallel AF$, and $BD \parallel AH$. | (Art. 531) |
| But | $AB \perp BC$ and $BD$. | (Art. 505) |
| $\therefore AB \perp AF$ and $AH$. | | (Art. 123) |
| $\therefore AB \perp$ plane $MN$ | | (Art. 509) |

                                             Q. E. D.

It is a concise style; yes. But ask any schoolboy of your acquaintance whether Euclid does not suffer constantly from

leaving too much to the imagination.  It is a style admirably
suited to a couple of aviators on a falling airplane, or a
subway passenger talking with a friend, but as a medium
of normal social intercourse it leaves much to be desired.

If Euclid were ever tempted to drop into poetry he would
probably have said " Love ‖ Truth," or " Man ⊥ the Stars."

And it all came about from the child Euclid's too intimate
association with his grandmother.  To realize what Complexes will do to a man's literary style we need only compare
the above citation from Euclid with almost any passage from
Mr. George Moore or Mr. James Branch Cabell's " Jurgen."
In these men we find a mellifluousness, an easy grace, a
rhythmic movement and a *savoir faire*, which testify plainly
to the absence of all inhibiting complexes.

# CHAPTER XXVIII

## THE ASTERISK AND THE DOTTED LINE

### (A Missing Chapter in Euclid)

Like nearly all literary and historical relics of the ancient world, the manuscript of Euclid's "Elements of Geometry" has come down to us in mutilated form.

Euclid, as is well known, was passionately interested in all kinds of points — points in a circle; points outside of a circle; points in a straight line; points outside of a straight line, etc. He was just as passionately interested in lines — straight lines; curved lines; broken lines, etc.; all of them conveying a more or less obvious meaning.

It seems odd, therefore, that Euclid should have omitted all consideration of what is unquestionably the most sexually significant of all points, namely, the Asterisk; and the most sexually significant of all lines, namely, the Dotted Line.

To the recovery of these lost phenomena in Euclid, the present author devoted himself. The search involved prolonged and exacting labor, but it was crowned with success. We are now in a better position to appreciate the full importance of Euclid the Elemental Amorist.

There is little doubt that the potentialities of the asterisk and the dotted line were first drawn to Euclid's attention by the contemporary writers of fiction. In the remorseless studies of life and sex to which the realistic novelists of Kos and the adjacent islands devoted themselves, there constantly arose situations which could be more safely imagined than described. That is where the symbolic line of asterisks or dots came in handy.

Euclid was no doubt familiar with passages in the realistic fiction of his time which looked as follows:

" It was only two hours since Fate, for its own purpose, had first introduced Glaucus to Sybilla.  Now that the festival lights were dying out and the moment had come to say farewell, he looked at her and said, ' Must you go? '

. . . . . . . . . . . . . . . . . . . . . . . . . . . . . . . . .

" The next morning dawned clear and crisply cold, with a lively breeze from the Phoenician coast."

There were realistic novelists who preferred the asterisk to the dot, but the general effect was the same:

" He looked at her and said ' Must you go? '

*   *   *   *   *   *   *   *   *   *   *   *   *   *

" The next morning dawned clear and crisply cold, with a lively breeze from the Phoenician coast."

There was a third school which made use neither of the asterisk nor the dot, but of the double space, but still with the same effect:

" He looked at her and said, ' Must you go? '


" The next morning dawned clear and crisply cold, with a lively breeze from the Phoenician coast."

It was argued in behalf of the double-space school of realistic fiction that it involved a considerable saving in linotype composition; but it was not a conclusive argument. As a matter of fact, the most successful realists of Euclid's time confined themselves to no one fixed style.  They made free use of the dotted line, the asterisk line, and the double space, as the occasion demanded and their fancy moved them. And both the occasion and the fancy operated with great frequency.

Nevertheless, it is a tribute to the genius of the Greek mind, that the realistic novelists were not long permitted to monopolize the advantages of the dotted or the asterisk style. The writers of clean, wholesome fiction for the home quickly adapted the device to their own use.

They began tentatively with a few dots or asterisks in the manner made familiar in our own time by Mr. H. G. Wells; as when he writes:

" Mr. Hengist picked up his paper. . . . It was the Morning Post. . . . He began to operate on his breakfast egg . . . thoughtfully . . . "

From this it was not a very long step to the full exploitation of the dotted and starred line. Greek best-sellers of the clean, wholesome type, began to show on every page passages like the following:

" Leuconoe smiled as she glanced at the vegetable pedler kneeling over his basket.

" I have seen better garlic in the market at three pounds to the drachma," she said. " Much more convincing garlic."

" The vegetable seller looked up at Leuconoe.

. . . . . . . . . . . . . . . . . . . . . . . . . . . . . . . . . . .

" Five minutes later Leuconoe in the kitchen was rinsing the garlic in cold water and singing to herself."

It is my firm belief that what the romantic school of Kos fiction achieved, our own writers of the clean and wholesome school will soon be practicing. As for instance:

" She looked up at him and smiled, with the tea-cup poised over the tray.

" Will it be lemon or cream? " she said.

. . . . . . . . . . . . . . . . . . . . . . . . . . . . . . . . . . .

" Five minutes later he asked for a third cup."

But, it may be asked, What has all this to do with Euclid? The answer is, A great deal. Euclid did not go in for fiction, but he did go in for geometry. And when his preoccupation with the Eternal Triangle began to manifest itself, the dotted line played a prominent part. To prove that two triangles

herb
roth.

The Asterisk

are equal if three sides of the one are equal respectively to
three sides of the other, Euclid was continually drawing
figures like the following:

## PROPOSITION  6

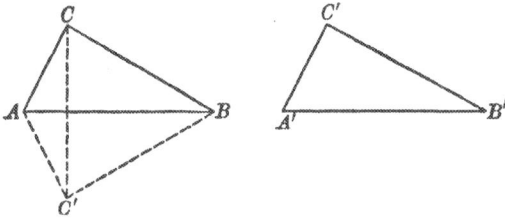

FIG. 8

Or take Proposition 12 in Book I.  A novelist would have
described the situation somewhat as follows:

" Right Triangle $ABC$ looked at Right Triangle $A'B'C'$
and $A'B'C'$ at $ABC$.  They knew that they were equal to
each other in respect to their hypotenuses and one of the
short sides.  But were they equal to each other in all respects?  They wondered.

. . . . . . . . . . . . . . . . . . . . . . . . .

" Yes, they were congruent in every respect."

Euclid, employing his own idiom, stated the same ·funda-
mental sex-truth in the following manner:

## PROPOSITION 12

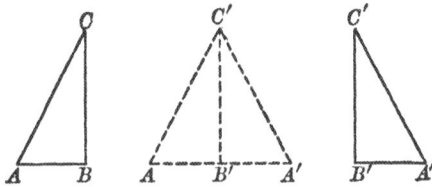

Fig. 9

Later mathematicians have developed the possibilities of the dotted line. I need only recall to my readers the familiar friend of their youth, the Binomial Theorem:

$$(X + Y)^{12} = X^{12} + 12\ x^{11}y + 66\ x^{10}y^2 \ldots\ldots\ldots\ldots\ldots 12\ x\,y^{11} + y^{12}$$

Among the mathematicians the dotted line is known as the Line of Infinite Suggestion.

BIBLIOGRAPHY

# BIBLIOGRAPHY

Maxwell, *Dream-Factors in the Fordney Tariff* (New York, 1922).

Pizzicati, *The Nightmare in Futurist Art* (Napoli, 1921).

Williams, *Sex-Elements in the Ferromanganese Industry* (Pittsburgh, 1920).

Dubost, *La Nevropathie du parallelogramme* (Paris, 1913).

Grove, *How to Bisexualize an Angle* (Annals of Academy of Science, Mathematical Sexion, 1922).

Jones, *The Labor-Day Complex at Grand Central Station* (New York, annually).

Jells, H. B. *An Outline of Hysteria;* original fortnightly serial edition, London, 1920; 2 vol. edition, New York, 1921; 1 vol. edition, New York, 1921; 4 vol. edition, New York, 1923; condensed edition, New York, 1922; expanded edition, New York, 1922; wide-margin edition, New York, 1922; uncut-edges edition, New York, 1922; Fox-Paramount film edition, with special titles, Hollywood, 1923; Sonorola edition, with record cabinet, New Jersey, 1922; Slampico edition, with foot-pump or electric attachment, New York, 1920; Radiophone WZQ edition, 1340 metres, New York, *passim; New Republic* edition, New York, 1921; *Snappy Stories* edition, New York, 1922; Incompletely-read edition, at various second-hand book stores.

Smith, *The Mirrors of Matteawan* (New York, 1923).

Horlick, *Musical Comedy and the Unconscious* (New York, annually).

Zumpft, *Die Mutterliebe und Schneidersrechnungen* (Maternal Love and Tailors' Bills, Erlangen, 1921).

Drinkhard, *Typical Cases from Canada and Cuba* (Publications of the Federal Warehouse Association, New York, 1922).

THE END

www.ingramcontent.com/pod-product-compliance
Lightning Source LLC
Chambersburg PA
CBHW020007290326

41935CB00007B/341